Isaac E. Taylor

Lupus of the Cervix Uteri and Female Genitalia

Isaac E. Taylor

Lupus of the Cervix Uteri and Female Genitalia

ISBN/EAN: 9783337239497

Printed in Europe, USA, Canada, Australia, Japan

Cover: Foto ©berggeist007 / pixelio.de

More available books at **www.hansebooks.com**

LUPUS

OF THE

CERVIX UTERI AND FEMALE GENITALIA.

BY

ISAAC E. TAYLOR, M. D.,

President of the New York State Medical Association; President of the Bellevue
Hospital Medical College, &c., &c.

WITH ILLUSTRATIONS.

Reprinted from the Transactions of the New York State Medical Association.

NEW YORK:

J. H. VAIL & CO.,

1888.

PRINTED BY

THE REPUBLICAN PRESS ASSOCIATION,

Concord, N. H.

LUPUS OF THE CERVIX UTERI, AND FEMALE GENITALIA.

I bear in remembrance that the statute of limitations as to time, as well as your own desire to reach the regular order of the business of the Association, invites me to an abbreviation of the remarks I propose tendering to you on Lupus of the Cervix Uteri, and the Genitalia of Women. I trust, in presenting the subject I have in view, that you will not be disappointed, and that it will impart some slight information and knowledge toward an acquaintance with a disease which is of every-day occurrence on the face, and which all of you have cognisance of in some of the varied forms which exist. It is very rarely recognised in the genitalia of women, and it is exceedingly rare when invading the uterus, cervix or body.

CASE I.

History.—I accompanied Mr. H., August 21st, 1885, to see his wife, consequent on a rather free uterine haemorrhage she had a few days before, and which required the attendance of a medical gentleman. It was the *first* and *only* flooding Mrs. H. had, and it was the *last*. She presented an anaemic appearance. In person rather fleshy, appetite good, and slept well. She had experienced no pain before the occurrence of the haemorrhage. It commenced suddenly. All the functions of the system were natural and well performed.

She married in early life, at the age of 17. Her confinements were quick and easy. Her recoveries were always favorable. She nursed all her children, and had six. All of her children had excellent health and good constitution, showing no constitutional taint of tuberculosis, externally or internally, in this respect resembling their parents. The grandchildren are perfectly healthy.

There was no history of cancerous disease of the uterus or elsewhere in Mrs. H.'s family.

Mr. H. died suddenly at the age of 73, from cardiac disease.

Vaginal examination. The uterus was found movable, presenting no hardness, no tenderness, no irritability. A soft feel was given to the touch on the cervix. No trace of blood from the examination. The vagina was much longer than ordinary, and natural to the feel. Its length was fully 3½ to 4 inches. The uterus was elevated, and could be felt externally above the pubes very distinctly.

The pudenda and vulva natural in appearance, though fleshy. The surface of the diseased part was a light pink color, elevated, uneven, and not uniform, with intervening spaces, covering the anterior and posterior part of the left side of the cervix. The edges were of a brighter color. When the cotton wool was brushed over it, a slight reddish discharge followed. The discharge from the diseased part was trifling, and of a light watery yellowish red color. There was no odor from the discharge. Dilute nitric acid was applied. Internally, Mur. Tinct. Ferri three times a day. Long narrow tampons of cotton wool were prepared to be used should flooding occur.

August 25, 1885. No haemorrhage; patient looking brighter and better; anaemic condition improved; very little discharge from the uterus, and of the same character, soiling only three or four napkins a day; bowels regular; appetite better; slept well; and had no pain.

A speculum examination was made with a large Fergusson instrument, having a bright sun-light. The cervix was brought clearly and distinctly into view. A slight oozing was seen from the ulceration. The ulceration covered half of the anterior and posterior part of the cervix on the left side, extending up to the fornix. The appearance would be properly termed tesselated, or pavement like. The drawing is a true copy (Plate 1).

Diagnosis.—Lupus Serpiginosus, or Tuberosus, of the cervix uteri.

The patient was visited every four days. Treatment continued till her return to the city, October 17th.

October 17th. Improved very much in every respect; discharge the same; no pain, and no odor. The ferruginous preparations were changed, and Donovan's Liquor and Fowler's Solution of Arsenic substituted, which were given alternately every month. During the winter, several of the more active escharotics were tried, such as pure nitric acid, chromic, lactic, and per nitrate of mercury. Every time these strong remedies were used they appeared to aggravate the disease. Curetting was not adopted for reasons which will be offered elsewhere. The dilute nitric acid was resumed January 4, 1886. This remedy was continued with the same internal ones till March 5, when the disease appeared to be improving. The posterior lip of the cervix showed signs of healing over. General health very much improved.

March 1st. Little if any discharge; nearly all the diseased surface healed over. During all this period of time—five months—there had been no pain, and no odor from the discharge. The patient had gained

Lupus Serpiginosis of Cervix Uteri.

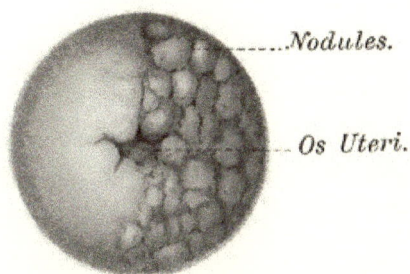

.....Nodules.

Os Uteri.

Mrs. H's Case, August 25, 1885.   *From Nature.—H. L. Shively.*

Lindner, Eddy & Clauss, Lith., N. Y.

in flesh, was stronger, and so much encouraged as to intimate that there might be no need for further medical attendance.

June 1, 1886. Returned to the country. Visited every week, continuing to mend. She visited her family, drove out, was bright, cheerful, and encouraged. The cervix almost all healed over.

Saw her but once during the month of July. *In statu quo.*

August 2d. For the first time complained of frequency of urination. On examination, the aspect of the pudenda, clitoris, and urethra appeared larger than natural, the meatus presenting a white, dull color, and firm, as well as the urethra; labia majora and minora increased in size, in length, and in breadth,—the right more than the left,—$3\frac{1}{2}$ inches long and $1\frac{1}{2}$ inches in breadth, and firm to the feel; the vagina not so dilatable; three small circular elevations on the superior part of the vagina, near each other; uterus movable, though it seemed to be larger; increase of discharge, and the ulceration commencing anew its destructive process; no pain and no odor; patient feeling very well; treatment the same; usual diet and Mariani wine ordered.

October 5th. Returned to the city. Since my last visit—September 7th—the patient had a malarious attack, which continued for ten days, materially affecting her. She had evidently become more enfeebled and depressed, and was not entirely relieved of it till November 2d.

From October 5th to October 25th—three weeks—she was too feeble to permit any examination. Only two were made between November 2d and 28th.

November 28th, 1886. The disease on the cervix had not progressed, although it had again commenced on the posterior lip, and had extended farther, presenting as originally the same elevated pavement-like appearance, divided into small oval and round spaces; discharge, no change; and no pain.

During this interval of time—nearly two months—the right labia had increased in length and breadth; clitoris longer, larger, and broader; vagina decidedly more contracted; the discharge same in quantity and color, no odor; bladder more irritable; appetite good; bowels regular; and slept well.

Rectal examination. Uterus in the same elevated position, and movable.

December 1st, 1886. Owing to the great increase and enlargement of the pudenda, vulva, and clitoris, the patient was not able to sit up, and she remained in bed.

January 1st, 1887. From November 2d the disease had been pursuing a more active and progressive course. New features, externally, of the most formidable character of the disease arose, assuming the hypertrophic form as well as the tuberculous or nodose. On the lower part of the right labia several small elevations, from the size of a pin's head to that of a pea, were recognised, and two or three were larger, assuming an oval shape.

January 6th, 1887. The measurement of the labia majora was 5¾ inches in length, 4¾ in breadth, and a further evolution of tubercles on the right labia and a few more on the left.

A few days before, the clitoris, which was very large, had increased, and small pea-like elevations were distinctly evident.

Previously, the clitoris presented a smooth shining surface, firm to the feel, and thickened. From this smooth glossy surface a slight watery discharge was oozing, and when this was wiped off the part would again weep. The appearance of the clitoris at this stage of the disease resembled the oozing tumor. The measurement was 1¾ inches in length and 1½ inches in breadth. No examination. Patient appeared very well.

At this visit my friend, Dr. J. W. S. Gouley, was present. A drawing was taken (see Plate 2). The drawing exhibits a true representation of the pudenda and clitoris.

The tubercles on the labia and clitoris had increased, and some tubercles had also evolved on the right groin.

Uterus movable; no increase since the last examination; no induration surrounding the uterus; several more circular and oval patches recognised through the speculum, although a much smaller one was used; the vagina losing its expansibility.

January 17th. Labia majora increased to 6½ inches in length, breadth 5 inches; clitoris much larger; ostium vaginae more contracted.

January 21st. My friend, Dr. C. A. Leale, who had been requested to attend Mrs. H. in case of any emergency which might possibly arise, was present at this visit. Patient the same as at previous visits in all relations.

January 27th. A clipping from one of the nodules was taken. The tubercles had increased in number, very much larger, more elevated, and of different shapes; labia not increased. The specimen was given to Dr. Hermann M. Biggs, Instructor in the Carnegie Laboratory, for investigation. No application could be made to the cervix. The vagina was washed out with tepid water, and the dilute nitric acid injected. This was continued on every visit, though the solution was made much weaker. It was simply for cleanliness.

February 17th. No change, except in the clitoris, which was becoming larger in every way, and nodose.

February 24th. Labia majora the same in size, 6½ inches long and 4¾ in breadth; clitoris larger, 2 inches in length, perfectly tesselated in appearance; patient becoming more feeble; evident signs of emaciation; pulse 100; sickness of stomach; tendency to diarrhoea.

March 1st. Failing; pulse feeble and small; in a semi-unconscious state; diarrhoea moderated; takes little nourishment.

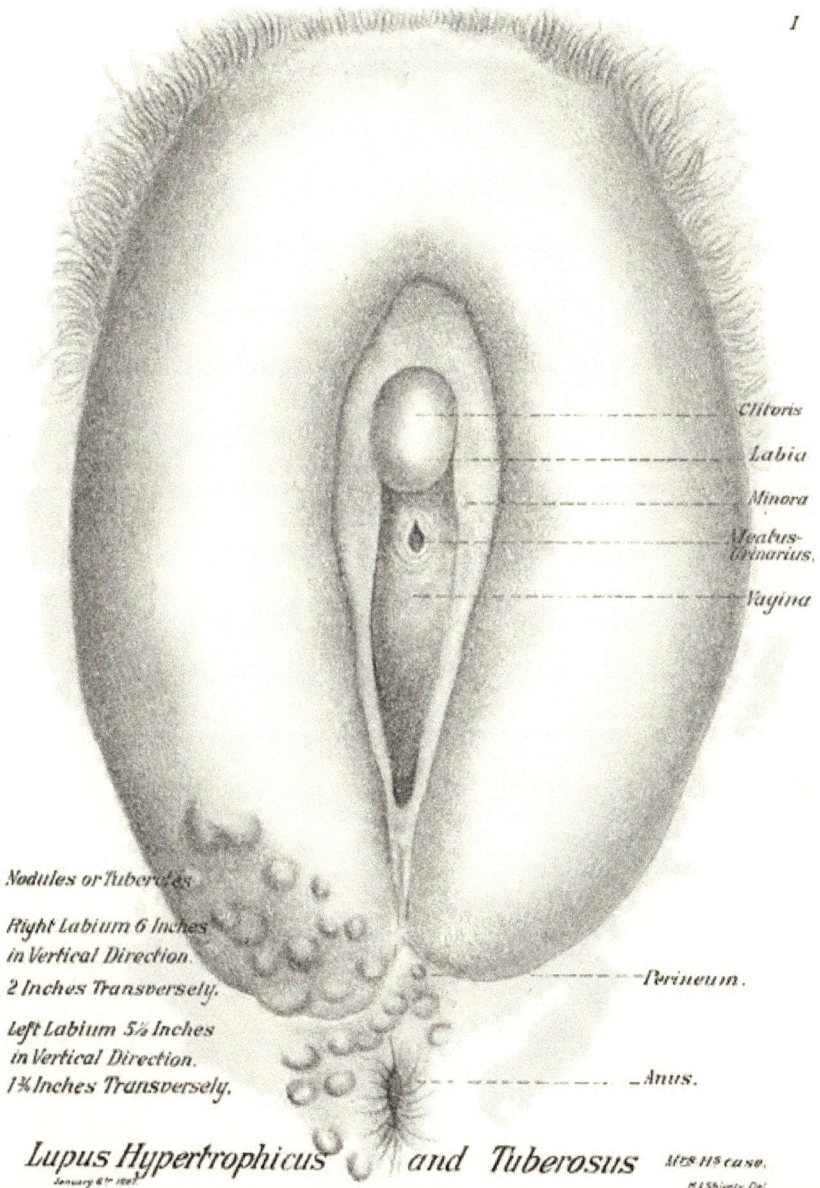

Clitoris

Labia

Minora

Meatus-
Urinarius.

Vagina

Nodules or Tubercles

Right Labium 6 Inches
in Vertical Direction.
2 Inches Transversely.

Left Labium 5½ Inches
in Vertical Direction.
1¾ Inches Transversely.

Perineum.

Anus.

Lupus Hypertrophicus and Tuberosus

Mrs HS case.

January 6th 1883

H.A Shively Del

II

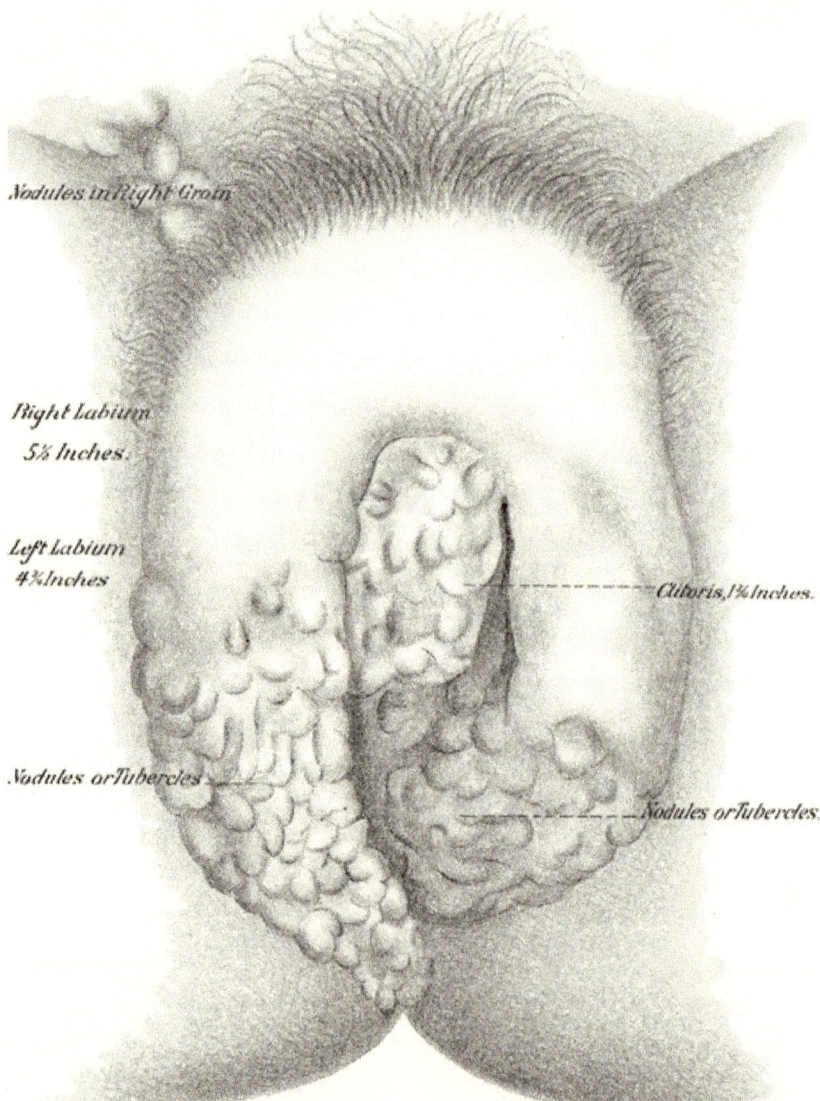

Nodules in Right Groin.

Right Labium.
5½ Inches.

Left Labium
4¾ Inches.

Clitoris, 1¾ Inches.

Nodules or Tubercles.

Nodules or Tubercles.

Lupus Hypertrophicus and Tuberosus

March 1ᵗ 1887

Mᶜ Hˢ case.
H L Shively Del

At this visit, on inspection of the pudenda, there was a marked shrink-age in the breadth and length, as also of the clitoris. There was a very distinct diminution of the tubercles and labia, when contrasted with the representation January 6th, 1887. The cast gives evidence of the great change (Plate 3).

March 4th. In Articulo Mortis. Died March 6th.

This exceedingly unique and rare illustration of Lupus Serpiginosus, commencing on the cervix uteri, presenting the same characteristic features as the same order on the genitalia of women, and afterwards the hypertrophic form developing on the pudenda and vulva, clitoris and urethra, corroborated the opinion which was entertained and expressed August 21, 1885, that the disease was one of the numerous forms of lupus, as described by dermatologists under different synonyms. I preferred the title Lupus, as the one most generally in vogue at the present day. Although the word "lupus," or "wolf-cancer," the ancient title, may not be considered as the most desirable in all its aspects, still it fulfils more than others the peculiar requirements incident to the nature of the disease, which may be viewed as a *sui generis* disease in almost every respect.

Different authors have included under such titles diseases which have no pathological relation to the affection from the commencement to the termination of the malady. Huguier selected the title, Herpes Esthiomène, simply because it gave prominence to one of its significant features, the tendency not only to destroy the tissues which were affected, but to reinvade a part which had healed over.

Noli me Tangere is another ancient title, which was applied, and still is, to the disease, because whatever local remedies were used, they were considered of no, or very little, value. The disease would progress at will to an unfavorable issue. The face is most generally the usual habitat, though occasionally it is seen on the extremities. All of you, I presume, have evidence of this disease in that locality, whether it is developed on the cuticular surface, or involving the mucous membrane of the mouth or throat, or the nasal orifices.

The tissues of the face which are affected are, it is true, different in structure from those on the genitalia of women, and from the uterus, in a physiological and anatomical aspect; still there is a close alliance and affinity between them.

It is immaterial, however, by what title we may elect to call the disease. From the distinct and peculiar action and nature of the destructive elements which it possesses, it proceeds steadily to advance, and accomplishes the death of the individual. The disease is one of the most formidable, stealthy, persistent, though slowly destructive in its nature, and no matter where it may be located it will outstrip all the skill of the most experienced and talented surgeon or physician eventually.

Cases of recovery are recorded, and I have evidence of its being cured. They may remain so for years, possibly to be again renewed. The occurrence on the genitalia of women, where the destruction of the vulvo-anal region is involved, is far more disagreeable and disgusting to the patient than when it has engaged the face. The deformity consequent on the enlargement and the length of the labia majora is in many instances excessive, and is one of its distinguished features on the genitalia.

The contraction of the vagina and rectum, which results sometimes after the ulceration has healed, nearly closes up these outlets by dense white, hard, and callous growths, or it may be from the cicatrisation which follows the destructive ulceration. At times the relation which naturally exists between the vagina and the rectum is perfectly destroyed, presenting a large chasm extending all around these outlets, freeing the rectum from its natural connection, and causing an extrusion of the rectum to a considerable extent, giving to these organs an appearance which is deplorable to behold. (See Plates 9 and 10.)

With all this mass of destruction, it is astonishing that, the disease going on internally and externally, the patient does not complain of any pain, or of very little, and seldom has there been any odor. If there is, it is faintly acid. The disease will go on *ad libitum* with its silent invasion and excavating process.

Frequency. Respecting its frequency or rarity on the pudenda, vulva, vagina, and uterus, Huguier reports seven cases on the pudenda and vulva. West, of London, had also seven cases. My own eight cases, including the present case, occurring *ab initio* on the cervix uteri, and afterwards manifesting itself on the pudenda and vulva.

On the contrary, Vidal says that in about 150 women affected with lupus vulgare, not one had the disease on the genitalia.

Kaposi had seen 1,200 cases of lupus vulgare, and in women as often as in men, but he had not once seen lupus of the nates and thighs extend to the labia. He had seen only one case on the penis. It is evident that there is a great discrepancy in the experience of medical men as to the frequency of the disease on the genitalia of women. It would seem to infer either an erroneous diagnosis on the part of the gynaecologists, or the non-presentation of these forms of the disease to other practitioners, and more especially to the dermatologists. Are the representations I present to you, so graphically drawn from nature, lupus? (See Plates 5, 6, etc.)

P. C. Huguier, of Paris, was the first, as a gynaecologist, to give emphasis to this disease occurring on the genitalia of women to the medical profession, in a memoir read before the National Academy of Medicine, February 8, 1848.

While attending physician to three large medical institutions at the same time, one of these being the New York City Dispensary, whose *clientèle* at that time, 1839, was very extensive, the class of diseases belonging to my department were diseases of females, and the non-syphilitic as well as syphilitic diseases of the skin. The service in those institutions claimed my attendance from 1839 to 1848, and since then the Bellevue and Charity hospitals for over twenty-five years. During that period of time, 1839 to 1848, nearly nine years, not a single case of lupus on the genitalia, in any of its various forms, was observed, as recognised since in the Bellevue and Charity hospitals and private practice. A number of serpiginous forms of syphilitic disease, with considerable enlargement of the labia, came under observation. The patients gave the history of, and had, constitutional taint. The appearance of this class of syphilitic disease sometimes closely resembles lupus.

In 1882 I presented a paper on some of the forms of lupus on the pudenda and vulva before the American Gynaecological Society. Graphic illustrations were produced from nature by Mr. Köhler, an experienced anatomical artist, and I now present them to you with a cast of Mrs. H.'s case, originating on the uterus. These representations embrace the serpiginous, the deeply ulcerated, the hypertrophic, and prominens, a form of tumor of the raspberry or strawberry appearance.

Matthews Duncan, in 1885, read two papers before the Obstetrical Society of London, one considering the hypertrophic and the other the ulcerative order of lupus, which he called lupus. The representations of several cases were chromo-lithographs. They present an entirely different appearance from those of which I have representations.

Several of the speakers, and among the principal ones was Mr. Jonathan Hutchinson, dissented from the opinion of Duncan, as well as the drawings, which were considered as representing syphilitic lupus. This view was coincided in by nearly all the speakers, except one who thought they were elephantiasis arabum. Others thought they were tertiary syphilis. In other words, it was either lupus grafted on a syphilitic constitution, or a constitutional syphilitic disease, with lupus intercurrent. I shall refer to this special point hereafter.

Mr. Thin, one of the most eminent and prominent microscopists in London, said that from his microscopical investigations they did not present the true development pathologically of lupus vulgare. He considered them to have a fibrous exudation, and that there was no evidence of cancer, epithelioma, or corroding ulcer, and that he should regard them as a disease *sui generis.*

It is only within a quarter of a century that an apparent approach to what might be conceded as a more correct opinion respecting the pathological nature of the disease has been entertained, though the most distinguished dermatologists and pathologists have given the results of their minute and careful attention, practically, as also microscopically, and it remains, *sub judice*, for farther investigation to ascertain if the disease has a specific bacillus belonging to it. The pathological disagreement, as also the difference respecting the title, the frequency, the cause, and the treatment, prevails as much at the present day as in past centuries.

The medical profession have established it as a cutaneous disease. The face is recognised as the most usual and frequent seat of the disease. We have positive evidence that it selects at times the delicate and soft structures of the uterus, pudenda, and the vulva-anal region, where various and distinct classes of the glandular system exist.

The difference in the nature of the disease may not appear as distinct as on the structure of the face and body. Similar relations we realise, however, prevail between the skin and the mucous membranes, and when the former is affected, papules which exist about the mouth externally will assume the character of aphthae or of vesicles when invading the mucous membrane.

Squamous diseases have sometimes attacked the mucous membrane, but instead of scabs and crusts we have only excoriations. In my own cases and those of the authors I have cited, the disease on the genitalia was not found as on the face, with scabs and crusts of the mucous membrane covering the cartilages or osseous structures of the nose, which constitute one of the characteristic features in many instances on the face. This may be accounted for, as the discharge emanates from a mucous membrane having different kinds of glands, which exist on the pudenda and vulva, the vagina and uterus, and the non-exposure of these parts to the atmosphere.

Pathology. Considering the results of the latest micro-scopical investigations, lupus is a series of affections conse-quent on a specific outgrowth of the cutis alone.

O. Weber, Berger, and others believed that it was a hyper-plasia of the rete malpighii with small round cells.

Huguier refers to Robin's and Lebert's results, who locate the disease in the same structures of the cutis. The cells were elliptical, and had all the characters of epidermic cells, each one being polygonal in form and more or less regular, but having a nucleolus.

Paget says that "The material scraped from the upper surface was formed of small epitheliform scales of various shapes, with nucleoli, and all were small and round." This examination was from a specimen of one of West's cases.

Wedl, of Vienna, differs from all these authorities, not only as to the seat of the disease, but as to its character and nature. He considers that a new formation of the connective tissue exists, and that the disease has its locality in the corium.

Auspitz and Kaposi coincide with the views of Wedl, as well as Klebs and Virchow. The small nodule or tubercle, when first recognised, is made up of small cells differing from the cells of the rete malpighii. The cells are sometimes ellip-tical, with short processes inclosing a round or oval nucleus with nucleoli. At times these cells equal in size those of a well marked fungus medullaris cell, and are more usually found in the soft raspberry-looking tumor of the prominent form.

My friend, Prof. J. W. S. Arnold, in a specimen taken from one of these tumors, recognised the giant cells. They were large, ovular, circular, irregular, and well defined. In the cases of lupus prominens, cell-heaps were recognised. These cell-heaps have been considered as peculiar to lupus. Giant cells, however, are found in scrofula, hyperplasia, syphilis, and tuberculosis. A small portion, which was easily removed from the cervix uteri, as well as a clipping from one of the tubercles on the right labia, in Mrs. H.'s case, was examined by Dr. Hermann M. Biggs, instructor in the Carnegie laboratory, and is thus described:

There was a moderate amount of hyperplasia of the malpighian layer of the skin, with at some points slight cellular prolongations of the inter-papillary promontories. The superficial layers of the corium contained here and there small collections of round cells, and a little deeper down were numerous irregularly outlined spaces in which lay loosely dense packed masses of epithelial cells. The tissue surrounding these spaces presented a very densely nucleated fibrous structure.

In some of these the masses of cells which formed the contents were so loosely placed that they had fallen out in the process of preparation.

The histological appearances were not unlike those sometimes found in recent secondary carcinomatous deposits in the skin. (Plate 11.)

Reference to the investigations of such prominent micros-copists and pathologists tends to show a material difference of opinion on these two essential points as to the nature of the disease.

Paget's and Biggs's observations go to show that possibly a carcinomatous element may prevail.

Wilson and Plumbe, of London, Lugol, Lulois, Doutrel-pont, Koch, and others, of France, regard lupus as a tubercu-losis of the skin, classed under the scrofulides.

Swimmer, of Buda Pesth, at a late meeting of the Berlin Congress, stated that the affection of the skin known as tuberculosis was a very rare one, while lupus occurred frequently.

Kaposi says that no clear conception of what is meant by tuberculosis or scrofula can be entertained, and he knows of no case of the disease in which scrofula was present at an early stage.

In none of my cases, nor in Huguier's, was there any external or internal evidence of a local or constitutional tuberculous disease. The manner of death in those who have died from the disease had no relation to tubercular affections in any way. I have no desire to ignore the fact, which is evident from the constitution in which scrofula or cancer or syphilis may exist, that there are no circumstances which would or would not predispose to their development, and be favorable to the production of the disease. The remarks of Swimmer and Kaposi corroborate what I have

previously stated, that it is a well established fact that
the generality of persons with tuberculosis or scrofulous disease, who are not constitutionally affected, seldom suffer or
have lupus. On the contrary, the constitution is seldom
undermined or broken down as in tuberculosis, but it continues in its integrity for months and years. Sometimes the
patient gains flesh, as Mrs. H. did at one time, and another
patient under my care who weighed over 200 pounds.

Leloir, in 1882, commenced the inoculation of lupus material. In one half of his experiments positive results were,
he says, obtained, in which a generalised tuberculosis occurred, and the tubercles were found to contain bacilli.

Twenty years previous to Leloir, in 1862, while practising
syphilisation in the Charity and Bellevue hospitals, several
essays were made to inoculate other patients. The result
was *nil*. Possibly my method was not the same as Leloir's,
though it was tried now and then for several months.

Koch has recently, it is reported, demonstrated the bacillus
tuberculosis four times in small numbers before he could be
certain of it. He examined at one time twenty-seven sections,
and at another forty-three sections, before finding a single
bacillus. He says he never saw more than a single bacillus
in a giant cell. Koch has obtained cultures of the bacillus
from blood stroma after the inoculation with material from a
specimen of hypertrophic lupus. There is no mention, however, whether any of these patients of Koch whom he inocu-
lated had tuberculosis.

Demme and Pfeiffer had found it constantly.

Mélassez, of France, looked in vain for it in several cases.
Cornil and Babes, having examined several sections from
each patient, found only one. I have not read of a case
recorded as yet where, in a typical case of lupus without a
constitutional taint, a bacillus was found. All the cases
thus far recorded were in a tuberculous constitution, or asso-
ciated with local diseases. We cannot doubt that lupus may
be grafted or occur in scrofulous, syphilitic, or cancerous
constitutions.

Instances have been recognised where syphilis has been cured, and lupus has afterwards declared itself, and gone on for years, and the constitutional disease has not presented itself again.

The original syphilitic taint may predispose to the formation of lupus, consequent on an impaired constitution, but it does not therefore impress on it the evidence which would demonstrate it to be of a secondary syphilitic nature. This may have been the case with Duncan's patients and others, but he says they did not give evidence historically of syphilis. Winternitz asserts that no case of the development of cancer out of lupus has ever been reported. Certainly none of the females I attended with lupus of the vulva had communicated the disease to their husbands, and this is the opinion of Huguier.

To what, then, does lupus owe the great destructive features which belong to it? Is it to the size and shape of the cells? We have seen that they are so various that from that circumstance alone they may have no special pathological peculiarity or significance as other tissues have. The want of a corresponding agreement respecting the bacillus tuberculosis is as divergent among the pathological investigators as the microscopical.

Has it been shown or proved at the present day that lupus has its own significant bacillus, as truly diagnostic of the disease as the bacillus tuberculosis in tuberculosis? Can it be asserted with verification that when the bacillus is recognised from a specimen of true uncomplicated lupus, because the tuberculosis bacillus is found, therefore lupus should be considered as a scrofulide? Could it be asserted also, that when lupus is seen in a syphilitic or cancerous patient, and a bacillus tuberculosis is discovered, therefore the disease is a scrofulide?

No disease has presented more difficulty of solution as to its true pathological nature, despite all the minute investigations in different ways to recognise the peculiar elements of destruction appertaining to it.

It has been asserted also by able and prominent dermatologists that there is no histological appearance that we can with propriety regard as significant of the disease. No! It stands at this day as a disease from its inception, and through all its different phases to its termination, as a disease *sui generis.*

In a clinical aspect, therefore, we have from this standpoint to look beyond the results of microscopical appearance, whether a bacillus exists or not, to the knowledge we possess respecting the cells, to whatever order or class they may belong, as the great destructive factors or agents; to their activity, persistency, stability, intractability, if not malignancy, in keeping alive the disease; and beyond this at the present time we have not gone. Commencing simply in appearance, as the disorder does, as a small nodule, and progressing stealthily outward and inward, performing in this way a dual office, we perceive the devastation going on for months and years,—the disease healing up at times, and supposed to be cured, but again renewing its former action of destruction. It is no matter what part of the genitalia it may elect as its seat: as long as these cells exist they will travel, and keep on their monotonous and subtile course as a disturbing element to the constitution till nature succumbs.

Lupus is credited with an alliance to some class of uterine disease, and more especially to corroding ulcer, epithelioma, sarcoma of the mucous membrane of the cervix, and tuberculosis.

Corroding ulcer is certainly an unusual form of disease of the uterus to meet with.

As far as some symptoms are concerned, an approach to an alliance may be entertained,—the duration of the disease, the mobility of the uterus, and its freedom from fixation to the surrounding structures, the pain not being so intense as in other uterine diseases. Commencing, as the disease does, on the cervix, yet in its nature and course there appears to to be very little or no similitude. The destructive process plays an entirely different part. In time the disease gradually makes its way to the structure of the cervix itself, and reaches the cavity of the body, excavating it, in fact shelling out the whole of the interior of the body, and leaving only a small part of the muscular structure of the uterus.

The discharge from the deep ulceration is sometimes considerable, and of an exceedingly foetid odor. Haemorrhages are not infrequent. The pain is compared to a burning-coal-like sensation. Death is most generally through a gradual wearing out of the physical constitution. Two cases of this affection have come under my observation; and I present to you a drawing of one of these cases (Plate 12) in an early stage, as a contrast to the one of lupus. One post mortem specimen was presented to the Pathological Society. With some of the prominent German physicians this disease has no identity.

Epithelioma has been referred to as having an alliance with lupus. Epithelioma is very frequent, commencing as it does also on the mucous membrane of the cervix in a superficial manner at first, slowly and gradually invading the whole cervix, destroying section after section, and propagating its deep destructive action to the surrounding tissues, amalgamating nearly all of them in one solid mass of cohesion and hardness. Haemorrhage is a frequent and prominent symptom, and sometimes a profuse serous discharge. The odor is especially significant. Pain becomes excessive, and continues with but few remissions in aggravated instances. There is but little respite for sleep; excessive emaciation, and the disease will seldom continue over two years.

Tuberculous disease of the uterus has been added to the list. This is an exceedingly rare disease of the uterus, and is principally confined to the body in its selection. The cervix may become similarly affected, and when it does the cervix becomes enlarged in consequence of the small nodules or tubercles. These tubercles are larger and more prominent, and distinguishable from the lupus nodule, which is flat and of different form. Those of tuberculosis are soft, consequent on the infiltration which they undergo, and they give issue

3

to a small quantity of thickish matter resembling pus. Eventually a rapid destruction ensues, and the patient dies through symptoms of constitutional disease incident to tuberculosis. The pus which is secreted from the serpiginous or ulcerated form of lupus is of a different nature and character, and is not regarded as a product of the breaking up of the newly organised substances, or of the normal elements of the tissues.

The pus which comes from the diseased surface of lupus, though small in quantity, is from a new formation out of, or from, a blastema. In most other diseases the morbific action affects only one tissue in particular, and it is more or less confined by its own specific action to such tissue. This does not hold in lupus. Its ravages have no respect to continuity of organisation. Having once destroyed the seat of its original development, it continues to nibble, mouse-like, its way through every tissue which appears in its progress, until its farther course is limited either by the destruction of the organ or by artificial means—an end which may be, and sometimes is, attained. No corresponding relation exists between the duration of the malady, or the destruction which it produces. No open ulceration took place in the nodules on the labia in Mrs. H.'s case, possibly because the time was short, only five months from the commencing evolution until death. The nodules commenced with a small elevation, increasing to the size of a pin-head, becoming larger, and by degrees reaching, some of them, to the size of a three-cent piece, assuming different shapes, round, elliptical, or oval, according to the pressure consequent on juxtaposition. The two drawings (Plates 2 and 3) presented of Mrs. H.'s case show the evolution of the tubercles, and the plaster of Paris cast which I presented to you exhibits it perfectly.

The long duration of the disease, as a general rule, though in some cases it may be a short time, rests in some measure on the nature and class of the disease and the constitution of the individual. The fact of its being cured at times, and not returning till after many years, and not proving fatal till after an interval of years unless some intercurrent disease supervenes, and not, as cancer does, in one or two years,— these are also some of its distinguishing features, clinically, beside those referred to. Where, may I ask, are the examples of ulcerating cancer being cured and relieved by treatment, as some of these cases have been? or where will we find cases of true cancer of the uterus or pudenda healed up for a while and renewed again? Where do you find the glandular system, after years have elapsed, take on a deep ulceration, or tuberculosis cured and manifesting itself again, divested of any constitutional affection?

From this essay of a rapid delineation of the diseases of the uterus referred to, there seems to be no direct alliance with them of lupus on the genitalia of women, from a clinical aspect or from microscopical investigations. Microscopic recognition, in connection with all that has been accomplished for a more correct and true interpretation of the disease, is certainly desirable, and the future may develop it. At the present time, however, we have to rest, clinically, on the pathological element, as we believe, of the *cells*, their great activity, ability, and persistency, and the diseased blood-vessels which sustain them, and to endeavor to divert them from their malignancy, when under treatment, as far as we can.

The treatment I refer to confines itself principally to the disease on the genitalia of women, for it is mainly and solely to the disease invading these organs that I have addressed myself. It is principally local, becoming sometimes surgical, depending entirely on the special character of the form to be treated.

Should the opinion be entertained that a constitutional element exists, those internal remedies which are generally approved of to modify that special condition are to be selected. Should the disease present the tumor-like or raspberry form, or the hypertrophic order on the labia become excessive, or long, white, hard excrescences arise, they can be removed.

The exsection may be either by the écraseur, crushing scissors, or galvanic cautery, and afterwards brushed over with the brown-heated cautery. The potential cautery it is not desirable to resort to, efficient as it may be, and justifiable. This will depend, however, on circumstances as to the nature of the case and locality. The location of the disease on the vulvo-anal region, where so much cellular tissue and the glandular element exists, ought to preclude its use, as sloughs have followed. The cicatrisation consequent on the suppuration arising therefrom would or might interrupt for a long time the healing process, and then it would have to be by granulation. An important and essential principle is to be borne in mind, which is, not to do injury to the healthy structures beneath and around the diseased part, if possible, as much harm has arisen from it.

The same principle is applicable to the stronger escharotics. Nearly all the active remedies of this class were tried in Mrs. H.'s case. They only aggravated the disease, and were discontinued, and a return made to the first remedy that was resorted to, the dilute nitric acid, which I esteem highly in nearly all these uterine diseases requiring a mild treatment in preference to other remedies. Cicatrisation was noticed as commencing, and afterward almost all the ulceration was healed over.

The milder escharotics are more usually applied in the serpiginous or ulcerative form. By the use of the milder escharotics on the principle I have stated, eventually a cicatrisation may ensue, or a restoration to an apparent cure, which results by a gradual absorption of the diseased tissues, a course which nature in her own way sometimes adopts, of repair through fatty degeneration instead of through the process of granulation.

CASE II. (Plate 4.) *Lupus Serpiginosus, or Extensive Superficial Lupus of the Whole of the Vulva; Hypertrophy of the Labia Minora and Majora.*
A. D., aged thirty years; married. Admitted into Bellevue hospital, April, 1863.
Constitution good. Lost her husband two years since. Menses regular; abdominal organs functionally normal. The disease commenced a

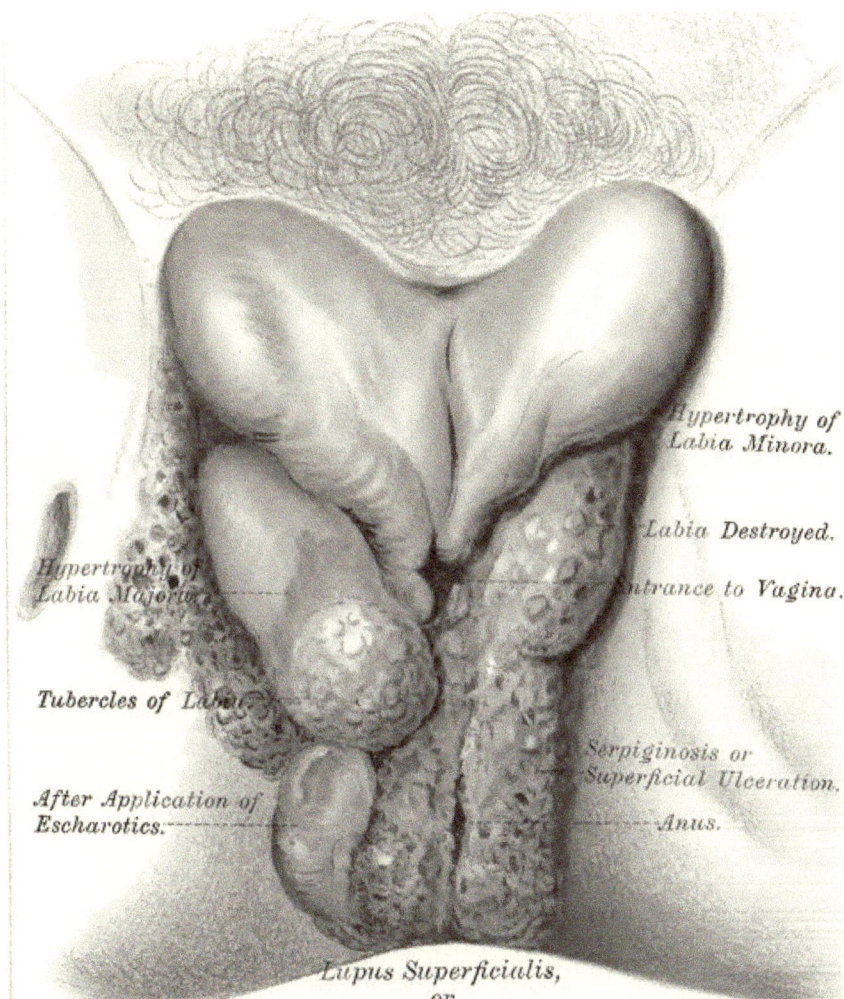

Hypertrophy of
Labia Minora.

Labia Destroyed.

Hypertrophy of
Labia Majora.

Entrance to Vagina.

Tubercles of Labia.

Serpiginosis or
Superficial Ulceration.

After Application of
Escharotics.

Anus.

Lupus Superficialis,
or
Lupus Serpiginosis.

A. D's Case,

year ago on the right and left side of the labia majora by a tumefaction of the labia. The lower end of the right labium was larger than the left, from which there was a slight viscid discharge, which excoriated the parts on the outside of the labia, at the junction of the labia and thigh. Both sides of the labia majora became affected to a considerable extent. On examination the labium majus, on the right side, was enlarged considerably, and the lower end knobbed, displaying the pea-like tubercles. The left labium was nearly all destroyed; the labia minora, right and left, were extensively hypertrophied in breadth, length, and thickness, covering the whole vulva like a pink apron, giving to the touch a firm, elastic sensation. The measurements were three inches in length, and one and one half in breadth; when they were elevated they presented the appearance of two butterfly wings. At the centre of the base was a slight fissure; at the lower part of the base of the flaps the orifice of the vagina was seen nearly closed, scarcely admitting the little finger. By some effort the index finger was introduced a short distance, and the disease was found to occupy the vagina to the extent of half an inch. On the right side, at the lower part of the labium majus, the spot of an inch in length and half an inch in breadth, presenting the smooth surface without any tubercular appearance, was produced by the chemical escharotics, which were applied a few times, which is also noticed in Case IV. (See Plates 7 and 8.)

The destructive ulceration commenced equally on both sides of the labia, extended downward, involving the perinaeum and anus; travelling upward, it occupied the lower part of the pubes. The discharge from this extensive surface was of a pale reddish, viscid nature, had but little odor, and was seldom very great. This is in some measure accounted for by the manner in which the pus is formed, coming from a blastema arising from the neoplasm or new formation of the connective tissues, and not from the breaking up of the normal elements of the tissue. The part from which the pus or fluid is produced is covered by a thin pellicle.

CASE III. (Plates 5 and 6.) *Lupus Prominens of the Tumor Form; Very Narrow and Thin Hypertrophy of the Labia Majora.*

M. S. consulted me in May, 1864, for small tumors, as she called them, on the vulva.

History.—This patient came from the western part of the state. Has been in the city for two or three weeks. Aged twenty-two. General health good; regular every month; menstruation first took place when she was fourteen. She has had no leucorrhoeal discharge; the general functions of the system are also regular. Parents in excellent health. The disease commenced several months ago by an itching of the labia; shortly afterward a redness appeared on both labia, which was not relieved by mild applications of sugar of lead, lime-water and milk, and other remedies locally. Three weeks after, small elevations on the skin were felt about the centre of the exterior of both of the labia majora. These elevations or pimples, as she supposed, gradually increased in two months not only in size, but also in extent and height; others manifested themselves lower down on both labia, and appeared to increase more than the others, until they reached the size which the plate represents.

On examination of the vulva, four vascular, scarlet-colored tumors, resembling large strawberries, were presented to view. The two upper, on both sides of the labia, were not so large as the lower ones. The right lower one was of an oval shape, two and one quarter inches long and one and three quarters inches broad, extending below the anus. The left was not so extensive. They were sessile.

The pea-like, tubercular appearance of the surface of the tumor was well marked, having white granular spots on it like millet seeds, which were produced by the sebaceous glands. She complained of no pain in them. The tumors were covered by a thin pellicle; there was no discharge of fluid from their surfaces.

On separating the labia a semicircular hymen was apparent. The whole appearance of the vulva was virginal. The hypertrophied labia were long and very thin, and extended four and one half inches; they were in close apposition. This form of hypertrophy is one of the marked peculiarities of this disease of the vulva.

Applications of the solution of the pernitrate of mercury were made, as well as strong acetic acid: no benefit resulted from them. On the contrary, in three weeks the growths had increased in size in all directions, and the upper and lower were approaching each other. Their bases were broader, the sebaceous glands larger, and the tubercular appearance very distinct. It was decided to remove them.

Before I refer to the treatment of this case, as Case IV is of the same order, I give a description of it, as it comes in appropriately at this time, in comparison.

CASE IV. (Plates 7 and 8.) *Lupus Prominens of the Tumor-like, Tubercular Form; Hypertrophy of the Labia Majora, Labia Minora, and Carunculæ Myrtiformes especially.*

H. B., twenty-five years of age; fair constitution, and generally good health. Menstruated at fifteen; unmarried; parents enjoying tolerably good health. Regular every month. She admits having had sexual intercourse, though seldom. When she consulted me the disease had existed over a year. She had had occasionally leucorrhoea, but no other disease on the vulva. The disease commenced by a small excrescence, as she termed it, on both sides of the labia, and was soon followed by others. Various local applications were advised for her comfort and to overcome the disease, but they were of little benefit. She experienced no pain from them, but their presence annoyed and distressed her very much. There was only a slight discharge from the tumors.

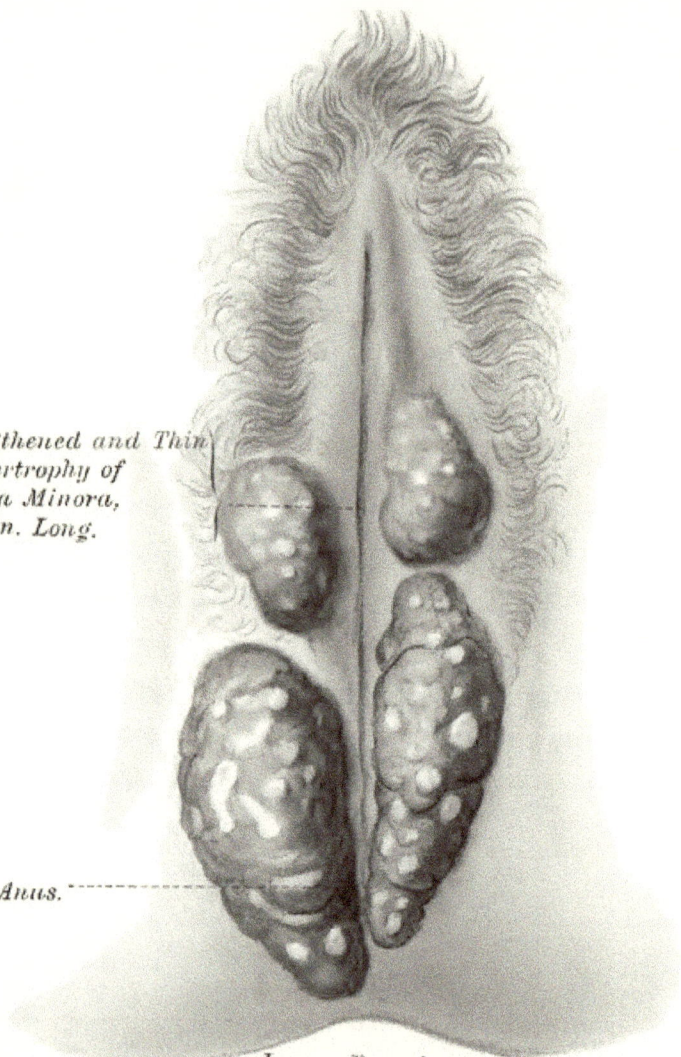

Lengthened and Thin
Hypertrophy of
Labia Minora,
$4\frac{1}{2}$ in. Long.

Anus.

Lupus Prominens.

M. S's Case, 1864.

Drawn from Nature, by Robert Kohler.

Lindner, Eddy & Clauss, Lith., N. Y.

Case IV No. 2.

Two weeks after No. 1

Lengthened
Hypertrophy
of Labia Minora,
4½ in. Long.

Anus.

Lupus Prominens,

S's Case, 1864.   Drawn from Nature, by Robert Kohler.

Lindner, Eddy & Clauss, Lith., N. Y.

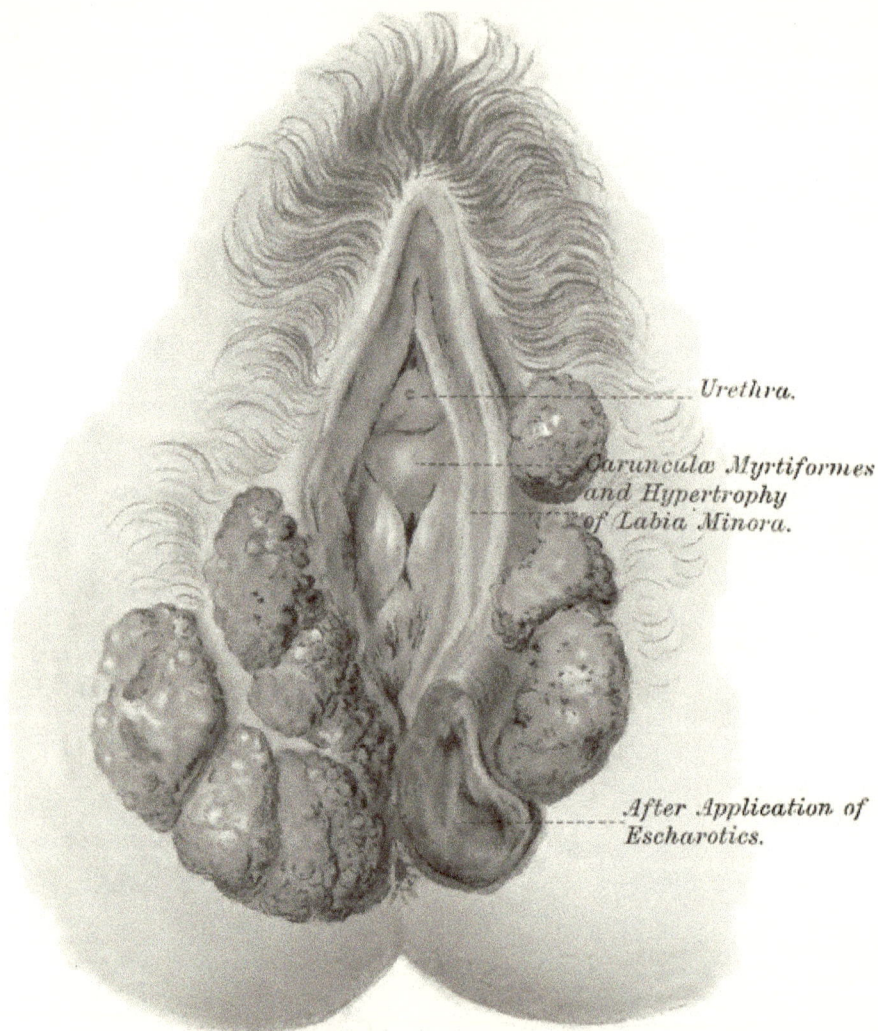

Urethra.

Carunculæ Myrtiformes
and Hypertrophy
of Labia Minora.

After Application of
Escharotics.

Lupus Prominens,
or
Tumor like Lupus of the Vulva.

H. B's Case, 1864.   Drawn from Nature, by Robert Kohler.

Lindner, Eddy & Clauss, Lith., N. Y.

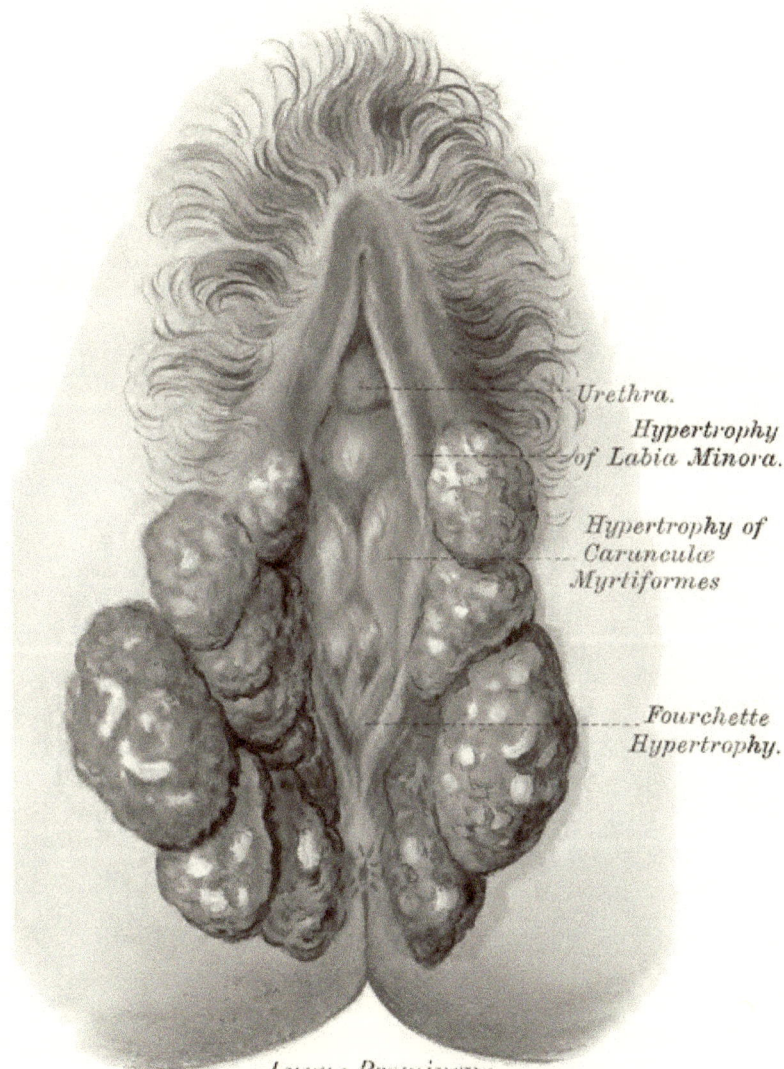

Case V, No. 2.
Three weeks after No. 1

Urethra.

Hypertrophy
of Labia Minora.

Hypertrophy of
Carunculæ
Myrtiformes

Fourchette
Hypertrophy.

Lupus Prominens,
or
Tumor like Lupus of the Vulva.

H. B's Case, 1864.   Drawn from Nature, by Robert Kohler.

Lindner, Eddy & Clauss, Lith., N. Y.

On inspecting the vulva, the external labia, as well as the internal, were moderately hypertrophied in length; the carunculae myrtiformes considerably so. The labia were everted, exposing the caruncles. There were eleven scarlet-colored, strawberry, prominent-looking tumors,— seven, of different sizes, on the right labium, externally, each separated from the others, having the pea-like or tubercular character of the disease, with small sebaceous nodules on all of them. On the left labium, externally, there were four of the same nature. The tumors on the right side extended down below the anus, and were in juxtaposition with those on the left. The tumors were of different forms and sizes

The different agents which were used for three weeks were of no value. During that time the disease had increased, and presented a more extensive appearance. The seven on the right labium had increased so much in height, breadth, and length as to measure on the surface three inches in length and one and three quarters in breadth. The tumors appeared to overlap each other, and had a solid base; the sebaceous glands were more developed; the lower one, which presented the smooth appearance, was produced by the application; the discharge was no more than at first.

The treatment of both these cases, so perfectly allied in appearance to each other, was similar, by the local application of the mineral escharotics, and once by the cautery. It was apparent that no good would arise from this treatment.

Amputation of the growths was decided upon. The patients were admitted into Bellevue hospital,—one in 1864, and the other in 1865. A few days later the tumors were separately removed by the crushing scissors or écraseur; Case IV before the class. No haemorrhage followed the operation. After their removal the bases of the tumors were brushed over with the brown-red heated cautery, and lotions of biborate of soda and acetate of lead were ordered to be made *pro re nata.*

In three weeks both patients were discharged. Cicatrisation had taken place, with the expectation that they might remain cured. The constitutional treatment was continued.

Case II, four months afterward, was free from any return of the disease. I saw her several times. The internal treatment was discontinued.

In Case IV, although the patient was seen occasionally during three months, the disease showed evidence of returning on the right labium. The local treatment by the brown-red cautery kept it from increasing for six weeks, when the woman appeared quite well; and two weeks afterward no return was noticed. The white cicatrisation presented no evidence of any change favorable to a return of the disease.

CASE V. (Plates 9 and 10.) *Extensive Hypertrophy of the Clitoris, and Labia Minora in length; Destruction of the Vulvo-Anal Septum; Dissection of the Cellular Tissue all around the Rectum, with Procidentia Recti.*

S. H. was admitted into the Charity hospital in the spring of 1865. Forty-one years of age; appearance haggard and distressed; emaciated, having had diarrhœa for some two weeks. Had had three children; labor natural, but tedious; children of ordinary weight. Had been irregular for the last two years. First labor, perinaeum lacerated to the anus. Denied ever having had syphilis. There was no evidence of any secondary or tertiary symptoms; had no nocturnal pains in the head or limbs; no cutaneous disease; no affection of the mouth or throat. The disease had existed for over two years; it was only within the last few weeks the rectum became procident. Had no severe pain.

The appearance of the external organs was very formidable, with the rectum external fully three inches. There was no eversion; it was a complete destruction of the cellular tissue all around its natural attachment. The anus was not affected; the labia majora are extensively lengthened and hypertrophied; the clitoris remarkably thickened and enlarged; measured three inches in length and one and three quarters in breadth. The labia minora were slightly enlarged, firm and solid to the touch, semi-cartilaginous in consistency, and of a dull white color, as were the labia majora; the whole of the vulva, when the clitoris was depending, was covered by it and perfectly smooth. When it was elevated, as the plate will show, and expanded, it presented a semi-circular appearance, showing its relation to the labia minora. At the lower end of the right labium minus were noticed four large tubercles, and one below. Lower down was another, where the ulceration existed, as well as on the cellular tissue of the external part of the rectum. The vagino-rectal septum was perfectly destroyed; the orifice of the vagina is perceptible in the diagram where the clitoris is elevated.

The patient was made as comfortable as the nature of the case would admit, by an effort to retain the rectum within, and to arrest the diarrhœa, and by good nourishment. Death took place in two weeks. No post mortem.

The three remaining cases were of the superficial order, and the treatment carried out as mentioned in my general remarks. In one case it involved the whole of the vulva and pubes, and reached the posterior perinaeum. The patient was under observation over two years, and gave no sign of syphilis during all that time. She weighed over two hundred pounds; never suffered any pain, but was very much annoyed and distressed by the great trouble she experienced in giving attention to the application on the parts, and by the discharge, which was not very great. She would leave the hospital after remaining in it three or four months, materially improved, and sometimes, to all appearance, healed over. In the course of two or three months more she would return again. This was the course the disease pursued repeatedly during the period I had charge of the patient.

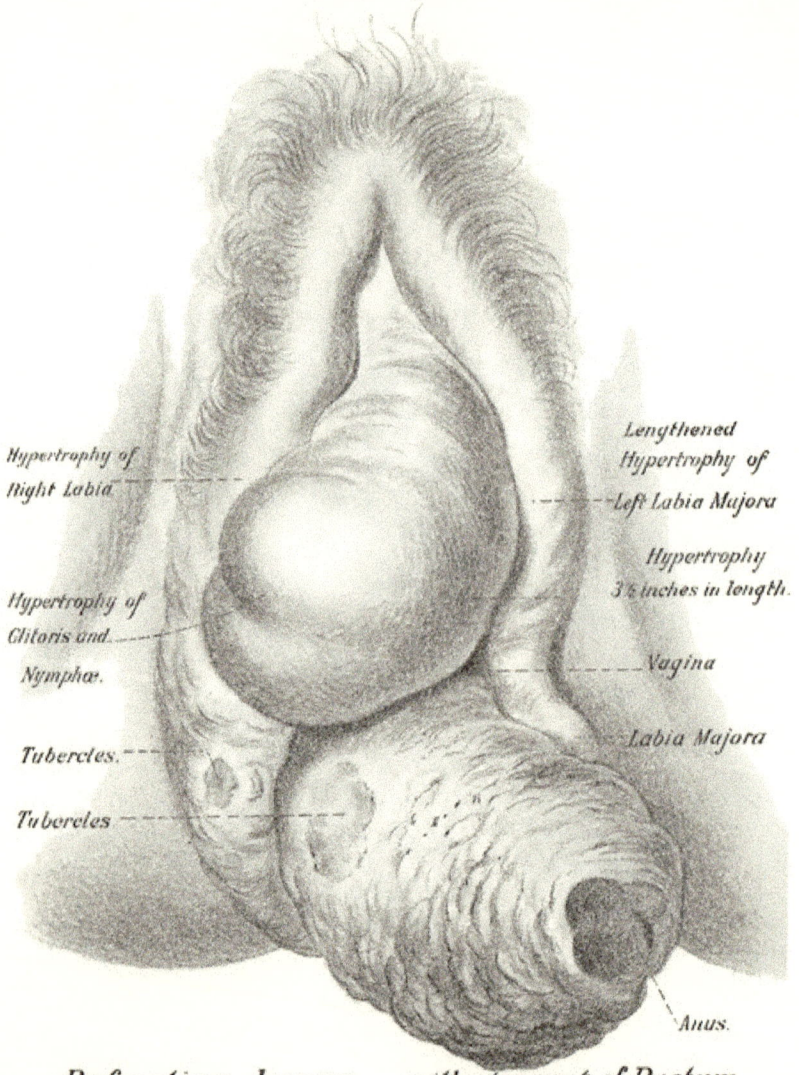

Case VI. Nº1

Hypertrophy of
Right Labia

Hypertrophy of
Clitoris and
Nymphæ.

Tubercles.

Tubercles

Lengthened
Hypertrophy of
Left Labia Majora

Hypertrophy
3½ inches in length.

Vagina

Labia Majora

Anus.

Perforating Lupus. with descent of Rectum.
S.H. 1865.   Drawn from Nature, by Robert Kohler.

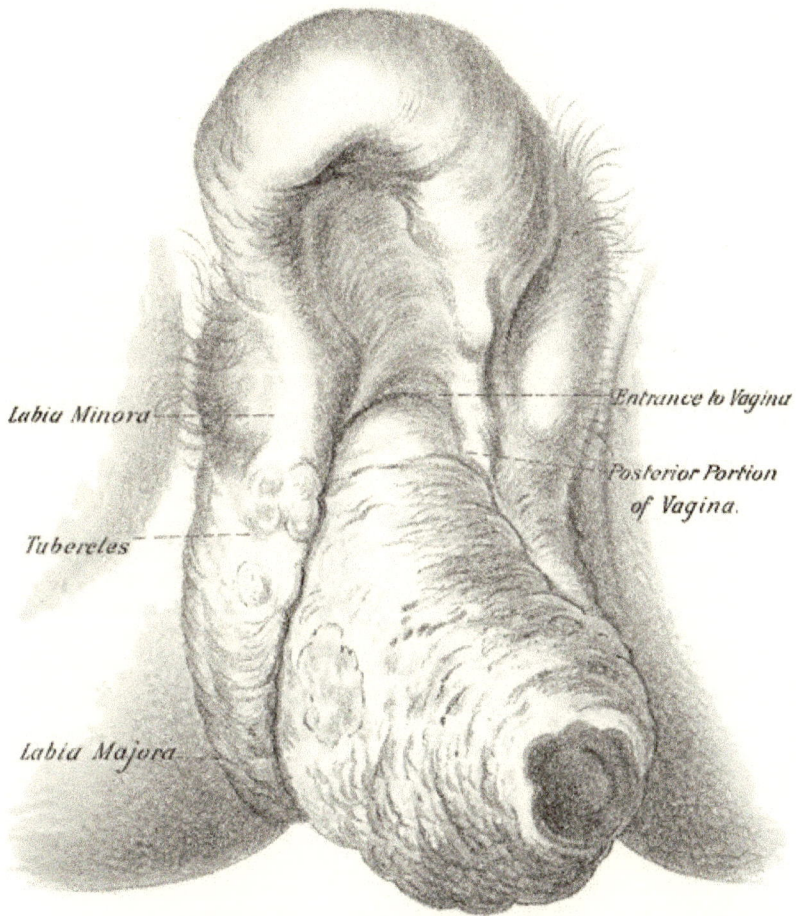

Case VI. No. 2

Labia Minora — — — — — — — — — — — — Entrance to Vagina

Posterior Portion
of Vagina.

Tubercles — — — — —

Labia Majora — — — — —

Perforating Lupus.
with descent of Rectum

S.H. 1865. Case.   Drawn from Nature, by Robert Kohler.

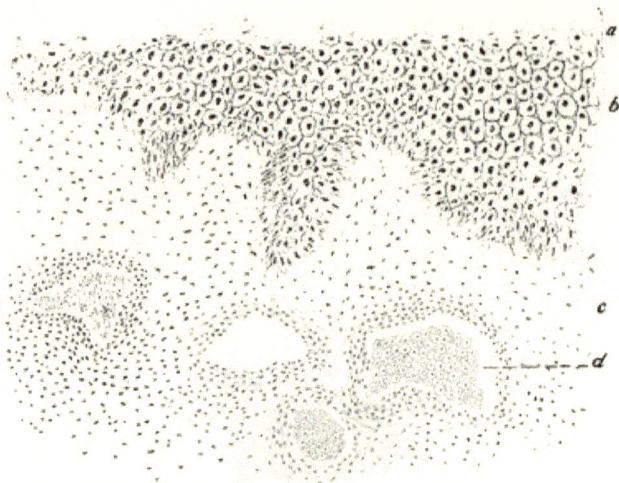

Section through Lupus Tubercle.

X 250.

Mrs. H? case.   H.M. Biggs
                                 after nature.

(*a*.) Superficial laminæ of the epidermis.

(*b*.) Rete Malphighu showing slight hypuplasia.

(*c* and *d*.) Corimu and sub-cutaneous tissue, with deeply uncleated,
          round-celled tissue below forming spaces, which contain masses
          of epethelioid cells.

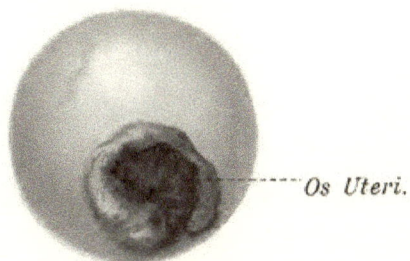

Corroding Ulcer.
Commencing.

------ *Os Uteri.*

Mrs. S's Case, 1863.　　　　　*Drawn from Nature, by Robert Kohler.*

Lindner, Eddy & Clauss, Lith., N. Y.